DAYS

BOOKS BY PAUL BOWLES

NOVELS

The Sheltering Sky

Let It Come Down

The Spider's House

Up Above the World

NOVELLA

Too Far from Home

SHORT STORIES

The Delicate Prey

A Hundred Camels in the Courtyard

The Time of Friendship

Pages from Cold Point and Other Stories

Things Gone and Things Still Here

A Distant Episode

Midnight Mass and Other Stories

Call at Corazón and Other Stories

Collected Stories, 1939–1976

Unwelcome Words

A Thousand Days for Mokhtar

AUTOBIOGRAPHY

Without Stopping

Days: A Tangier Diary

LETTERS

In Touch: The Letters of Paul Bowles (edited by Jeffrey Miller)

POETRY

Two Poems

Scenes

The Thicket of Spring

Next to Nothing: Collected Poems, 1926–1977

NONFICTION, TRAVEL, ESSAYS, MISCELLANEOUS

Yallah! (written by Paul Bowles, photographs by Peter W. Haeberlinb)

Their Heads Are Green and Their Hands Are Blue

Points in Time

Paul Bowles: Photographs (edited by Simon Bischoff)

DAYS

A Tangier Diary

Paul Bowles

AN **ecco** BOOK

HARPER  PERENNIAL

NEW YORK • LONDON • TORONTO • SYDNEY

HARPER ● PERENNIAL

First Harper Perennial edition published 2006.

The Library of Congress has catalogued the Ecco Press paperback edition as follows:
Bowles, Paul.
   Days : Tangier journal : 1987–1989 / Paul Bowles.— 1st paperback ed.
      p. cm.
ISBN 0-88001-282-X
1. Bowles, Paul—Diaries. 2. Bowles, Paul—Homes and haunts—Morocco—Tangier. 3. Tangier (Morocco)—Social life and customs. 4. Authors, American—Twentieth century—Diaries. 5. Composers—United States—Diaries. I. Title.
PS3552.0874Z465   1992   818'.5403—dc20
[B] 91-36440 CIP

ISBN-10: 0-06-113736-7 (pbk.)
ISBN-13: 978-0-06-113736-5 (pbk.)

06 07 08 09 10 RRD 10 9 8 7 6 5 4 3 2 1

CONTENTS

ILLUSTRATIONS

Paul Bowles Portrait, Tangier, 1986

Paul Bowles and Mohammed Mrabet, Café Hafa,
Tangier, 1983

Rodrigo Rey Rosa, Tangier, 1988

Phillip Ramey, Café Vienne, Tangier, 1989

Buffie Johnson, New York, 1989

Abdelwahab El Abdellaoui, Tangier, 1988

Abdelouahaid Boulaich, Tangier, 1986

PREFACE

Three years ago Daniel Halpern wrote me asking if I kept a diary. I replied that I did not and never had, not seeing any reason for engaging in such an activity. He wrote again, suggesting that I start one immediately, since he would like to include whatever resulted in an issue of *Antaeus* to be devoted only to diaries, journals, and notebooks. I told him that I thought the result would be devoid of interest, since I would have nothing to report. All he wanted, he responded, was a record of daily life in today's Tangier. I agreed to try and did what I could with the project, although I was not very faithful, often allowing two weeks or more to elapse without writing anything. What went on during the periods of silence I have no idea, but doubtless the unrecorded days were even more humdrum than the others.

I suppose the point of publishing such a document is to demonstrate the way in which the hours of a day can as satisfactorily be filled with trivia as with important events.

Paul Bowles
Tangier, Morocco
August 1990

DAYS

August 19, 1987

Clear. Walked to Merkala. The *cherqi* was violent, and raised mountains of dust along the way. On the beach hundreds of small children, hardly any adults. The boys were beating each other with long strips of seaweed. Constant smell of the sewage coming out of the conduit at the east end of the beach. Lalla Fatima Zohra was right to forbid the public to use the place a few years ago. But that was during the cholera epidemic. A letter from Paris saying that Quai Voltaire will not agree to letting me inspect the galleys of any book they may publish. I never asked to see galleys. I wanted to see typescript before it was set up in type. They called my request *"légalisme excessif."* Buffie found her two thousand dollars and passport, hidden somewhere in the flat.

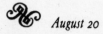 *August 20*

Paid my last visit to the consulate, and was given the copy of my will, which they'd been keeping for me. Moving vans in front of the Résidence. Afternoon brought a M. Jebari, doing a thesis at the Sorbonne. His first suggestion, *La vie et l'oeuvre de Paul Bowles,* turned down. When he called it *L'horreur et la violence dans l'oeuvre de Paul Bowles,* they accepted it. Ridiculous. Claude Thomas came by, resentful of the new contracts Quai Voltaire have sent her to sign. I hope she doesn't eventually lose patience with them and refuse to translate any further works. I count heavily on her for *Up Above the World.* Bourgois writes that he expects her to take on the volume of Jane's letters.

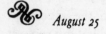 *August 25*

Curious how difficult it is to sustain anger, once the initial flush of it is over. For three days L. has been coming here to spend the entire afternoon. Twice or three times a year he arrives from Boston, where he's busy writing that biography which I rejected before he started. Weidenfeld is aware that it's unauthorized, and I've repeatedly told L. that I won't help him in any way. At least he asks no questions. Conversation with him is like talking with the doctor immediately after he tells you: "Yes, you have cancer," and then goes on: "But let's speak about something else." I wonder if he knows how deeply I resent his flouting my wishes. Probably not, since I say nothing, show nothing, and after all this time, even feel nothing.

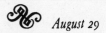 *August 29*

L. said good-bye this afternoon; he leaves tomorrow, certainly no more advanced in the preparations for his project than when he came. During the six afternoons he spent here Mrabet did almost all the talking. I think L. must be better equipped now to write on Mohammed Mrabet than on anyone else.

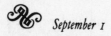 *September 1*

Jane's agent in New York tells me that the French
Society of Authors refuses to pay me any royalties
on works by Jane unless I can furnish documents
proving that I am her legitimate heir. It's *In the
Summer House* which has precipitated the trouble.
Plaisirs Paisibles, being only a book, went off without a murmur. But the play was broadcast. The
society apparently considers radio and TV as need-
ing stricter controls: more money involved.

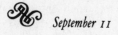 *September 11*

Finally saw Mrabet's villa-cum-stable at Mraierh.
(Jerez calls the place Charchumbo.) I was startled to
see that he'd combined the two in one edifice. Jean-
Bernard, who was with us, thought it natural, said
it was a common arrangement in France. Of course
it is here, too, in the more distant regions. But
animals bellow and bleat and smell and draw flies.
I can't believe it will be inhabitable. However much
enthusiasm Jerez may feel, she won't live in it long.

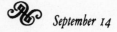 *September 14*

I looked through *Libération*'s questionnaire of two years ago: *Pourquoi écrivez-vous?*—this time to see what was the most usual answer. Very few writers claimed financial necessity as a reason for exercising their profession. Many admitted that they had no idea why they wrote. But the majority responded by implying that they were impelled to write by some inner force which could not be denied. The more scrupulous of these did not hesitate to admit that their principal satisfaction was in feeling that they were leaving a part of themselves behind—in other words, writing was felt to confer a certain minimal immortality. This would have been understandable earlier in the century when it was assumed that life on the planet would continue indefinitely. Now that the prognosis is doubtful, the desire to leave a trace behind seems absurd. Even if the human species manages to survive for another hundred years, it's unlikely that a book written in 1990 will mean much to anyone happening to open it in 2090, if indeed he is capable of reading at all.

October 3

Yesterday two men from the Wafa Bank called on me, handing me a letter from Casa, asking that I lend them two small drawings by Yacoubi for an exhibit they intend to hold there later this month. I said I had no Yacoubi drawings—only paintings— and they answered that drawings had been specified. Instead of shrugging and saying: *"Je regrette beaucoup,"* I added that I had had drawings, but that they had fallen behind the bookcases in one room or another, and I didn't know which room or which bookcases, and that I had no intention of moving those heavily laden objects in order to search. A bad idea, since they both volunteered to empty the shelves. Several thousand books. They're coming back this afternoon to do that work. In the meantime I've spoken with Abdelouahaid and Mrabet, both of whom advised me not to let them start. It would take several hours in any case. But A. and M. were in agreement that if the drawings were found and borrowed, I'd never see them

Paul Bowles Portrait, Tangier, 1986
(Credit: Cherie Nutting)

again. So now I must face the two Wafa men and say
the thing is impossible.

An even more unpleasant prospect is having
the British TV crew and *animateurs* arrive week
after next to do that interview. This I dread more
than anything because of my disappearing voice.
(Buffie insists I have cancer of the larynx, and has
no patience with me because I won't go for X rays.)

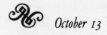 *October 13*

"When a Jew is dead, he's dead," said Gertrude
Stein. Yet both she and Alice Toklas were bad Jews.
Stein was a secret Christian Scientist; Toklas openly
embraced the Roman Catholic faith in her later
years. Is this regression?

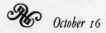 *October 16*

Three and a half decades ago Saïd Kouch, Jane's
Arabic professor, said to her: *"Tous les agréments
de Tanger ont disparu."* It was true then, and

meaningless now. Whatever charms the town once had have long since been forgotten. Bulldozers have run wild over the countryside, vegetation has been hacked away, and trees everywhere chopped down. Nothing surprising about that. Suburbs have to be put somewhere. But the housewives of the fancy new villas in these suburbs scatter their refuse from the windows, and send their servants to the empty lot next door, to add to the mounds of garbage already there. Buffie has gone back to New York.

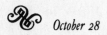 *October 28*

Last night the television crew returned to London, after eleven days in a Morocco without sun. They didn't mind the clouds and rain in Tangier because this made a good accompaniment to *Let It Come Down,* but they had hoped for sun in Fez, and above all in the Tafilelt, where they meant to film landscapes of sand dunes. But the desert was wet and gray. (Today the sky is cloudless.)

 *October 31*

Twenty or more women and girls, apparently on
their way to a wedding, walk along the street
pounding on drums. Behind them a dozen boys
follow, pitching stones at them. Hostility between
the sexes begins early.

(They were not on their way to a wedding;
they are all seated now on the ground at the top of
the hill opposite my bedroom window.)

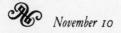 *November 10*

Smihi has been shooting *The Big Mirror* for four
weeks. Last night he came to remind me that he
wanted Mrabet and me to be in a bar scene, not as
extras, he was careful to add; he wanted us to con-
verse. I reminded him that such a dialogue would
have to be in Spanish. I suggested that Jerez de la
Frontera should replace Mrabet, who could be sitting with other Moroccans. Smihi agreed, and said
he'd call her hotel when he left here. He told me
to be at the Palace today at 2:00 P.M. I went. Jerez

was there, in evening dress. The Palace was locked. It was cold in the street. We waited until 3:30. Mrabet drove us back here. At 4:00 Jerez and I returned (on foot) to find the nightclub still shut, and no actors or crew in evidence. At 5:30 Gavin drove us home. Some time after 6:00 Smihi arrived, apologetic. Just the same, he tried to get me to go back with him to the Palace. I refused, he left, and Jerez and I ate dinner.

The men from the Wafa Bank came and carted away a large Yacoubi painting for their exhibit in Casablanca. "That's the end of that picture," said Mrabet.

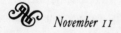 *November 11*

Mrabet says he had only to look inside the Palace (which he did at some point in the evening) to know that he would never have agreed to be in such a scene. The Moroccans sitting there were all of the kind that drink alcohol and consort with prostitutes, and he could not afford to have himself photographed in their company.

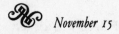 *November 15*

Rodrigo and I were in the Fez Market yesterday.
He drew my attention to a tray of mushrooms at a
vegetable stall. "These look exactly like what we
call San Isidros," he said. San Isidros are psilocybin
mushrooms, he went on, in case I didn't know,
which I didn't. He was excited to think they grew
in Morocco. I thought that if such a drug existed
here, people would know about it, and they very
clearly don't. But he bought a dirham's worth and
went home, saying he was going to brew them in a
tea. Today he came in triumphant. "They're the
same hongos. The same thing as in Guatemala."
The brew, which he said had a disgusting flavor,
kept him awake all night, writing rather than hal-
lucinating. It's hard to believe that psilocybin is sold
here in the market and that no one is aware of it.
This is probably because mushrooms are not a part
of the diet of Moroccans. Still, the Europeans who
buy them must have had some strange and unex-
plained experiences.

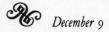 *December 9*

A relatively quiet period after the protracted frustration of last week, when it took six days of running between the postal authorities, the customs, and the censors to get the galleys of *Call at Corazon* into my hands. Owen is publishing a new volume of stories using that title.

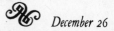 *December 26*

The most ridiculous gadget of the year, in the show window of an Indian shop on the Boulevard Pasteur: a deodorant stick with a built-in compass.

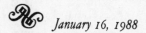 *January 16, 1988*

I have a document from the Société des Auteurs et Compositeurs Dramatiques telling me that I have been admitted *"comme membre succession."* Presumably I may now receive those royalties, although who knows? The French are still French; like the Moroccans, they part with money only under duress.

 *January 17*

Buffie writes that she was questioned by the police in connection with the murder of Donald Windham. She suggested that the aggressor perhaps had taken exception to something he had written.

 *January 20*

Every morning, weather permitting, I set out on a long walk. It's supposed to help my leg. It makes no difference whether it does or not; I go anyway. Each day I walk to the inaudible accompaniment of a different popular song. It's not necessary to look for them; they pop up from my unconscious. It was some time last summer when I realized that all these old songs from the 1920s were there. I can never remember what song it was that preoccupied me on the preceding day, so now I write them down. Today it was "Red Hot Mama."

Paul Bowles and Mohammed Mrabet, Café Hafa, Tangier, 1983 *(Credit: Mary Ellen Mark)*

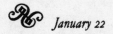 *January 22*

Today's song: "I Gave You Up Just Before You Threw Me Down." I seem to be weak on my legs, or so Abdelouahaid tells me, holding me up. But the weather has been colder than I remember its ever being in Tangier, and that is inclined to keep one's muscles from responding immediately.

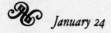 *January 24*

"Sleepy Time Gal." Rodrigo returned yesterday, having had to go twice to Panama on his way to and from Guatemala.

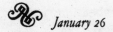 *January 26*

"Sueños de Opio." Each time I read an article about what the journalists call "the tragedy of Sri Lanka" I wait for the finger of blame to be pointed at the English. Incredibly, it never is. Instead, race, religion, and cultural tradition are combed over until the final impression the reader comes away with is that the conflict was inevitable. Everyone knows that the Tamils did not emigrate to Ceylon on their own initiative. Why did the British want them there? Because they needed an impoverished, helpless group of agricultural workers who could be forced to work for minimal wages. The Sinhalese could not be forced; the Tamils, being in alien territory, were at the mercy of the British.

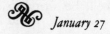 *January 27*

"The Alcoholic Blues." The report of Donald Windham's murder proves to have been a hoax carried out by someone who had a grudge against Buffie Johnson. She blames it on a student who lives on the West Coast. He must be the one, she says, because he has AIDS and "can't wait to make trouble" for others.

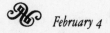 *February 4*

Three Japanese called on me today: a Mr. and Mrs. Inuhiko and Riki Suzuki, the editor of the literary monthly *Shincho,* published in Tokyo. Subject of discussion: their plan to translate and publish certain texts of mine. They want me to agree to furnish a preface "for the Japanese public." I told them how *The Delicate Prey* had been pirated twenty-five years ago in Tokyo and published under the title *Kayowaki Ejiki.* This may have been impolitic. Mr. Suzuki said he did not want to deal with an agent. So we'll be working without a contract.

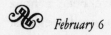 *February 6*

Rodrigo has bought a falcon. When Mrabet heard this, he decided that he was going to get it away from him, and began to announce his plans for teaching it to hunt.

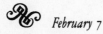 *February 7*

The weather is so bitterly cold that I've abandoned my walks. Mrabet arrives early and makes a big fire in the fireplace. I get up to find it roaring. Wonderful these mornings when the temperature in my bedroom is 38° F. Rodrigo has the bird in a cage. He says it's gentle and seems to have no fear of him. It eats fresh raw beef.

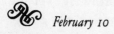 *February 10*

Three Italian journalists all afternoon.

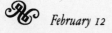 *February 12*

Brazilian journalist with intriguing first name of Leda. Rodrigo brought the falcon here. A beautiful bird. R. wants to take it to the top of the mountain and set it free. "So it can eat people's chickens," says Mrabet.

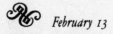 *February 13*

Abdelouahaid and I drove with Rodrigo and the
cage to the high point above Mediouna. There was
a hard climb over sharp rocks to get up there. Abdelouahaid helped me. When the cage had been
opened and the falcon had been persuaded to come
out, Rodrigo threw it upward into the air, and we
stood watching it as it flew. There was a strong *cherqi*
blowing which seemed to keep it from rising very
far. It flew straight toward the northwest over the
pine forest, as though it knew where it was going.
Little by little it went up. By that time the cold had
got to my bones. I came home and got into bed.

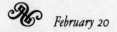 *February 20*

Incapacitated with a cold since the day we took the
falcon up to the mountain. L., my persevering biog-
rapher, has arranged a concert of my music as part
of the Manca Music Festival in Nice for the third of
April. Hard to be properly annoyed with him when
he goes to such lengths to be agreeable. He even

offered to come from Boston to Tangier and fetch me, if I'd go. I shan't go. I'm too old to put up with being stared at.

 March 2

Went to the Wafa Bank to inquire about the Yacoubi painting. It's in Casablanca; the exposition won't be held until April or May. Abdelouahaid and Mrabet believe that the exposition is fictional and that the painting has been sold. This seems unlikely.

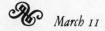 *March 11*

The house which Mrabet was supposed to be building for Jerez is finished, he says, but he is not going to let her have it. The eighty thousand dirhams she gave him a year ago before she returned to New York he used, irresponsibly and one might say criminally, to build the stables he wanted for his animals. The house itself, over the stables, cost three or four times as much, money which I unenthusiastically supplied after the big scene with Jerez. When she came back, he expected her to bring more cash, but she didn't, so that he felt obliged to feign great anger. (Attack, before you can be accused.) The house is not yours, he told her. He then proceeded to furnish it, saying he was going to live there himself. There's not a chance of that, since the house has no water, electricity, or drainage. Neither he nor Jerez could live there. As soon as fair weather sets in I'll go out to see the place. Now that it's finished I'm curious.

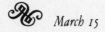 *March 15*

A disappointing version of *The Sheltering Sky* arrived today from Madrid. Alfaguara used the same title as the Buenos Aires edition of 1954: *El Cielo Protector.* A careless translation marred by omissions and errors. Too bad.

Rodrigo Réy Rosa, Tangier, 1988
(Credit: Cherie Nutting)

April 19

No one was certain whether today or tomorrow would be the first day of Ramadan. We knew only last night when the sirens sounded that today would begin the fast. (This is the second Ramadan with sirens instead of cannon. *Allez demander pourquoi.*) One shot, and you were over the boundary in the land where all is forbidden. They tell you that with today's traffic the cannon would not be heard. This may be true at sunset, but at half past four in the morning the city is silent. Strange that no Muslim has spoken of the ludicrousness of using an air-raid siren to herald a holy day of fasting.

Every year I have to remember to warn people who come for tea that they must leave well before sunset. The hour directly after that is the time to be inside, out of the street. It's the favorite hour for attacking foreigners. The streets are absolutely empty. Not a car, not a pedestrian, not a policeman in view, everywhere in the city. One of my guests, an elderly American woman, was knocked down, kicked, and robbed in the street in front of the apartment house. I felt vaguely guilty of living in a place where such things are taken more or less for granted. But the real guilt is that which I feel in the presence of Muslims. They are suffering and I am not. Here at home I'm obliged to eat and drink in front of them. They always claim that it doesn't affect them to see someone eating. If I want to eat, I can eat, they say. There's no one telling me I can't eat. This is true, but the social pressure is such that anyone seen to be eating in public is arrested and jailed.

Thirst is more painful than hunger, they say. Smokers are irascible for the first few days. As the month wears on, skirmishes between individuals increase in number. But no one will admit that he is short-tempered because of Ramadan. Says Abdelouahaid: "If you're going to be in a bad humor because it's Ramadan, your Ramadan has no value, and it would be better not to fast at all." Nevertheless they *are* likely to be fractious, and I take care not to contradict or criticize them.

April 24

I have a spider whose behavior mystifies me. It's
the kind of spider with tiny body and very long
legs, and it spins no web. It spends its days hang-
ing by one filament from the bottom of a marble
shelf behind the door. For the past three weeks it
has been going every night to hang four feet away,
near the washbasin. When morning comes it returns to its corner. There are no insects for it to
catch at either location, but it never misses a night.
If I let anyone know of its existence it's sure to be
killed. Spiders are not encouraged to live in the
house. Rahma is such a poor housekeeper that
the spider probably can count on months of pri-
vacy. If Mrabet or Abdelouahaid should catch
sight of it, they would unthinkingly crush it. I
don't know why I assume that it's entirely harm-
less, except that it looks nothing like the spiders
that attack. These have heavier bodies and thicker
legs, and are intensely, militantly black.

April 25

Went last Friday to Mraierh. The climb was painful then, but now after three days the pain is worse, and in the upper leg. This with the usual fire in the *mollet* makes it very hard to walk. I have to assume that eventually it will be better.

 April 27

After a week or so of springlike weather, we've now gone back to January, with covered sky and *cherqi,* and showers from time to time. I've noticed that it's next to impossible for Tangier to reach a warm temperature without the arrival of that cursed east wind, which immediately makes everything feel twenty degrees colder. This is why July often seems more glacial than a calm day in midwinter. I've gone back to having a fire in the fireplace. Pain still strong in my leg, but less so than yesterday, most of which I spent in bed. I'm using a combination of Adalgur and Alpha-Kadol. My mistrust of "patent medicines" dates from early childhood, when I heard only denunciations of pharmaceutical products bearing trade names. They were all reputed to be poisonous. Now it seems that pharmacies (in this third-world country, at least, and probably in most other places as well) sell nothing but such products. It is impossible to get a druggist to go into the back room and prepare a medicine with his hands. Here this may be a blessing; God knows what fatal errors are thus avoided.

 May 3

Typical tale of Ramadan violence at the market of Casabarata. A man who prepared *chibaqia* was sitting on the ground, hoping to attract buyers. (*Chibaqia* used to be made with honey; now, there being no more honey, it's made with sugar syrup, and isn't very good.) Another man carrying a little portable counter of combs, pocket mirrors, toothpaste, and similar objects, sat down near the first, who immediately ordered him to go somewhere else. The second man said he was going to sit there only for a minute, because he was tired. Then he would go on. The *chibaqia* seller roared: "Safi!," whipped out a long knife, and slashed the other with a downward motion, severing his jugular. The wounded man rose, took a few steps, and collapsed. His four-year-old son stood watching while he bled to death. This was the second killing at Casabarata since Ramadan began two weeks ago. There have been others, in other parts of the town, but I didn't get eyewitness reports of them as I did of this one.

 May 4

Jerez off to New York today. Unfortunately she came yesterday, bringing a big bunch of roses and lilies. Friday Mrabet had bought an armful of white roses from Kif Kunti, and had arranged them in a large vase. It was Jerez's fatal idea to put Mrabet's roses into a smaller receptacle in order to make room for her own more spectacular array. I didn't think Mrabet would be pleased to see this, but I wasn't prepared for his exaggerated reaction. The insults came fast and thick. Each time she tried to speak, he shouted louder. Anyone used to living here during Ramadan would have backed down and given up trying to reason with the adversary, but Jerez seemed to think conditions were normal, and continued to ask if she had ever done anything to harm him. His shouting grew louder; the insults came in Arabic, Spanish, and English. Then he began to hurl cushions at her, and finally hauled off and gave her a resounding crack in the face. Jerez was bending over him, so she did not fall. But Mrabet jumped up, seized a log from the fireplace, and swung at her, to hit her on top of the head. My shouting at him to sit down and shut up had no effect, but Abdelwahab, who was here as well as Abdelouahaid, came between them and calmed Mrabet for a moment. (Abdelwahab is a Riffian, so that Mrabet was more inclined to listen.) But then Mrabet must have felt that he had been bought too easily, and began to bellow that he was in a room full of Jews who should be killed and not allowed to pollute the air breathed by a Muslim. With this he left the room, and we heard him continuing his

Phillip Ramey, Café Vienne, Tangier, 1989
(Credit: Cherie Nutting)

insults and obscenities as he banged around the kitchen. Jerez by this time was sobbing, and Abdelwahab decided to leave, which he did so quickly that he left his umbrella behind. Abdelouahaid merely sat, shaking his head. He whispered to me: "A horrible man. Heart of tar." I think he was shocked by Mrabet's behavior. I was not shocked, having seen other instances of his insensate fury, but I was ashamed that all this should have happened in my flat, and to a guest of mine. When Jerez went out, still weeping, he shouted: "If you come back from New York, I'm going to kill you!" Five minutes earlier she had whispered to me: "Do you think he'll kill me?" and I had smiled and said: "Of course not." So his parting shot was not calculated to comfort her. It's some consolation to know that when she returns it will no longer be Ramadan.

Before Mrabet went home he excused himself to me for his outburst, saying: "She wants to drive me crazy. She kept saying I was a thief. Can she prove it? Does she have a witness?" It's pretty absurd to consider that all this was ostensibly about a bunch of roses that got put into the wrong receptacle, or so it would have seemed to an on-looker. In reality Mrabet has a bad conscience, and when a Moroccan feels guilty, he attacks.

 May 5

The spider, after having been absent for the better part of a week, has suddenly decided to return to its regular nocturnal haunt, where it stays the whole time, day and night. It seems to me there's something suspect about this. The identical spot where it used to spend its nights, yet I'm not convinced that it's the same insect. It looks smaller and feebler than before. If it's a different individual, what has happened to the original, and why does this one hang exactly in the place where that one hung? An entomologist could probably give a completely unexpected and satisfying explanation.

May 6

Rodrigo left last night for a week's trip through the
south. Since he's never been to Tinerhir I suggested
he go there from Marrakech and continue eastward
to Er Rachidia and then north to Midelt. He seemed
determined to go over the Tizi N'Test to Tarou-
dant, which he already knows. If he does that, he'll
probably not go to Tinerhir.

The cassette of the concert in Nice last month,
which I thought was lost because it had been
removed from the envelope, turned out to have
merely been confiscated by the censors because it
was not declared. Abdelouahaid brought it to me
this afternoon. The two-piano works were execra-
ble—worse than I'd expected. Some of the songs
not bad. The solo pianist managed, by dint of rushing like a cyclone through everything, to hit more
wrong notes than right ones. Why won't pianists
look at tempo markings? My suspicion is that they
imagine they make more of an impression by play-
ing as fast as they can, regardless of the metronome

indication, like typists eager to show how many words per minute they can turn out.

 May 7

From time to time when we're driving in the country, Abdelouahaid recounts something that happened or is said to have happened in a village we're passing through. Some of the stories are of the sort that it would never occur to me to invent. Others are banal, like this one. A couple facing increasingly hard times in their *tchar*. Last few chickens die of an epidemic. If we want to go on living, we'd better leave now while we can still walk. With the girl enceinte, they start walking along the trails, and come to Bab Taza at night. A man sees them and realizes they are from the country, asks them if he can help them. The girl says: "We're looking for a place." "A house?" "Yes." "Come. I'll show you a good house." Takes them to a house he has just bought with the intention of selling it. Before they go in the husband asks the price. (He is entirely without a guirch.) The house is completely empty. After they have been shown around, they ask if they may spend the night in it, and give the owner the reply in the morning. He agrees. They bid one another good night, and the owner leaves them there. The husband goes to the well to draw water for washing and taking supper. Sees a small wooden box floating in the shadows down there. When he brings it up, it is locked. He and his wife decide that the owner knows nothing about it. They open it. Full of banknotes. In the morning when the owner comes, they agree to buy the house, which they do with half the amount in the box. Abdelouahaid loves stories about hidden treasure, which are invariably without interest.

 May 16

The one enjoyable attribute of Ramadan was the *rhaïta* solo played in the minaret of each mosque at the times of the call to prayer. This year they have

Buffie Johnson, New York, 1989
(Credit: Cherie Nutting)

done away with music. I suppose someone came up with the idea that it was anachronistic or unorthodox. "People don't want to listen to somebody blowing a *rhaïta,* anyway," says Abdelouahaid. "They have music on the television." In 1977 I recorded the oboe concerts nightly for the entire month of Ramadan. Unconsciously I must have suspected that sooner or later they would dispense with them. Good things do not continue.

 June 20

Very little to write about. I've been receiving clippings in various languages, all of them announcing Bertolucci's intention of filming *The Sheltering Sky.* But in the cinema world any statement can be construed as propaganda, so I still have no idea as to whether or not he'll make the movie. People find it hard to believe that Helen Strauss included no time-limit clause in the contract when she sold the film rights back in the fifties. So if Bertolucci has acquired them, I don't know from whom.

June 26

The hernia has been giving me too much pain too
unremittingly for me to go on with it. Dr. Rawa
agrees to remove it if I'll let him do it with a local
anesthetic, and if I'll leave the hospital immediately
afterward. This seems to me the ideal way of han-
dling it. The Hôspital Kortobi has a very unsavory
reputation. The less time one spends in it, the safer
one is.

 July 3

A dark day. Phillip Ramey volunteered to go to the
hospital with me and wait during the operation.
Abdelouahaid was in the car outside. He hadn't
believed that I'd be able to walk from the front
entrance to the gate, but I made it without trouble.
The anesthetic began to wear off as we drove home,
and the rain came pouring down. The deluge
managed to cut off the electricity, so that when
we got to Itesa, the lift was not running. By this time
I was not very clear in the head. Someone came

upstairs and fetched a chair from the flat. I was seated in it, and Phillip, Mrabet, and Abdelouahaid carried the chair and me all the way to the top floor. That part of the day is unreal. I was in bed; the ceiling suddenly sprang a leak and the cold rain began to drip onto my feet. The pain was bearable. But then Regina Weinreich, just in from New York, arrived to talk about the film she proposes to make for American television. It was not the right time to get into that, and I was not very happy. There are few things as unpleasant to look forward to as a prolonged bout with a television crew. In principle I agreed to it, but since it won't take place until October, I was thinking: Perhaps she or I will die first. That's one way of making the future a little more acceptable.

 July 23

I think of how nowadays I never go near the beach. Fifty years ago it was where I spent my summer days. The days when for one reason or another I did

not go, I felt were nonexistent, wasted. The Moroccans said I was crazy. Not even the men sunbathed
in those days. They believed the sun was poisonous.
After the war the younger men played football on
the beach, and now and then you saw a female
walking into the waves, heavily dressed, of course.
The Moroccan girl who lived next door to us in
Calle Maimouni got into the habit of taking the
women of the neighborhood down to the beach in
the afternoon. They would return before sunset
in great spirits. Of the girl, Jane said: "She's a revo-
lutionary. She's got the only pair of water wings in
Tangier."

 August 7

A delight to be able to walk as far as I please along
the waterfront without hernia pain. An inexplicable
official decree now provides the beach with two-
dozen special police whose job it is to keep people
not wearing bathing suits from walking on the
sand. Those who are dressed must remain on the

sidewalk. This apparently senseless new law works to the advantage of no one except the owners of the cafés and restaurants that provide cabins where bathers can undress. Said proprietors must discourage the traditional maneuver of undressing on the beach and leaving one's clothing behind during the swim. There is a good deal of thieving by a squad of young men who never cease to move among the piles of garments, removing watches, wallets, sweaters—whatever seems easy to carry. The additional police fail to deter these troublemakers, naturally. They all work toward a common end, and one wonders if perhaps the restaurant owners don't hire the hooligans along with the police to help enforce the new *dahir*.

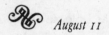 *August 11*

Yesterday an unfortunate day. Gavin Lambert and Phillip hatched a plan whereby we would go in two cars to Xauen, the two of them, Rodrigo, Krazy Kat, Abdelouahaid, and I. Before we started out,

Abdelwahab El Abdellaoui, Tangier, 1988
(Credit: Paul Bowles)

Gavin remarked facetiously: "Happy August the
Tenth" (the title of the first story Tennessee published in *Antaeus* back in 1971). It was the wrong
thing to say; in the story the tenth of August was not
a happy one for the protagonists. This August tenth
was hellishly hot. We in the Mustang were trying to
keep up with Gavin, who hurried ahead of us at an
unnecessarily high speed. At one point we saw his
car slow down and stop. We all got out and stood
uncomfortably in the strong sun. "It needs water,"
said Gavin. Abdelouahaid ran down to a river
below and brought up some water. This did no
good. Everyone got into the Mustang, sat one on
top of the other, and went to Tetuan, where we
drove from street to street in search of a garage
which might still be open. Since it was midday everything was in the act of shutting. Gavin had to stay
behind and find a *remorqueur* to tow him back to
Tangier. The rest of us continued to Xauen. When
we arrived it was just a little too late to get lunch at
the Parador. We had to satisfy ourselves with ome-
lettes and beer. It was even hotter up there than in
Tangier, for the lack of any wind. After lunch we

climbed up to Ras el-Ma and ordered tea. The bees were plentiful and insistent, and insisted on covering the rims of the glasses and sliding into the tea. Honey bees don't ordinarily sting, but there were so many here that there was nowhere one could touch one's lips to the glass in order to drink. Phillip took countless pictures and Krazy Kat made friends with everyone in sight. I wanted to get back to Tangier before dark, so we set out early. But with the extra people in the backseat, the car began to make agonized groans at each curve and pothole, so loud that it was hard to talk. The heat continued, because we were driving directly against the sun. Made Tangier just at dusk. End of Happy August the Tenth.

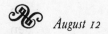 *August 12*

I was in bed last night having my supper, when Patricia Highsmith was ushered in. She'd been invited down from Switzerland to visit Buffie, but Buffie had forgotten she was arriving, and gone out.

I asked her to sit down, then told her where she could find some scotch and a glass, and we talked. After an hour or so she wanted to go down to Buffie's flat and unpack. I gave her a set of keys. Buffie returned as she was trying to unlock the door, and said: "I didn't realize you were coming," quickly adding, "today."

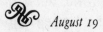 *August 19*

Robert Briatte says he intends to write my biography, and that Plon will publish it. He seems to believe that he needs no documents, which pleases me, since I have none to give him. He intends to give a certain amount of importance to the music, which also pleases me.

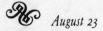 *August 23*

Drove Pat Highsmith and Rodrigo to Achaqar. We drank beer in the café they've built over the caves. Highsmith very agreeable company. Too bad she's leaving, although I'm afraid she hasn't had a very good time in Tangier. Buffie was ill most of the time and shut herself into her room, so that her guest was left to her own devices.

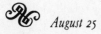 *August 25*

When Jerez came back from America she had her mother with her. She'd already rented a fairly large three-story house in Dar el-Baroud with fine views over the harbor, so they both moved in. Her mother seemed delighted with Tangier, and everything went well until Jerez (who had carefully refrained from admitting to her mother that the house in Mraierh was finished, but that she was not going to be able to have it) confessed that Mrabet was not going to give it to her. This news provoked a violent reaction. Jerez went off to Jajouka for a rest. When she returned with Bachir el-Attar, her mother refused to allow either of them to sleep in the house. It seems she confused Bachir with Mrabet, and thought it abject of Jerez to remain friendly

with a man who had swindled her. Jerez has begun
a campaign among those Americans interested in
the Master Musicians of Jajouka to help Bachir get
an American visa. I wrote a letter of recommenda-
tion to the consul in Rabat.

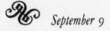 *September 9*

Jerez successful. Bachir given visa. They've left for
New York. Will they marry? The mother stays on
alone in the big house.

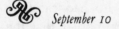 *September 10*

William Betsch arrived this afternoon from Paris,
bringing a particularly handsome (but heavy) art
book, *Cites d'Islam*. Why it's called *Cites* instead of
Cités I don't understand, nor can I find the word
in any dictionary. Betsch supplied many of the
Moroccan photographs for the book. He wants me
to write a preface for his own book *The Hakima*. I
agreed. A very strange amalgam of photos and text,

recounting what was probably the murder of a girl, but could have been suicide or a simple accident. A disturbing atmosphere suffuses the book. The lack of agreement among members of the family and those close to the girl makes all attempts to pierce the mystery impossible. It will be published by Aperture, which guarantees quality.

 October 7

Quai Voltaire has been trying to persuade me to go to Paris and appear on the television program *Apostrophes.* So far I've resisted successfully. But yesterday Daniel Rondeau flew down from Paris to try to make me see how overwhelmingly important such a broadcast would be for my "career." They would pay all expenses, yes, but the problem was in getting there. I didn't want to fly (when have I ever in the past forty years?); I could take the boat to Sète, and a car and chauffeur would pick me up there and speed me to Paris. Statistically, however, the auto routes in France are more lethal than the airways. And why more television? Especially when no payment is offered? Rondeau said that it was the duty of a writer to let his public see him, not to speak of the stimulus such an appearance gave to sales. I promised nothing, and he went back to Paris, after first commissioning an essay on the winter fifty-eight years ago when I lived on the Quai Voltaire. Can I manage even that?

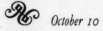 *October 10*

Lunch with Gavin Young, who advises me to go to Paris. He's on his way to Indonesia.

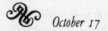 *October 17*

Decided it would do no harm to have a French visa in my passport, whether or not I make the trip. Called the always helpful Monsieur Bousquet, who offered to take me to the French Consulate and see that I was given a visa immediately.

Abdelouahaid Boulaich, Tangier, 1986
(Credit: John Claflin)

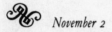 *November 2*

Wrote Rondeau today that I'd fly to Paris. He'll send me a return-trip ticket.

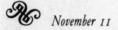 *November 11*

Knowing I've got to go is very depressing. Hard to think of anything else. I leave on Thursday, and am determined to return here Saturday.

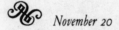 *November 20*

Got back last night. Acute euphoria going through the customs at the airport. Abdelouahaid had the Mustang outside, and Abdelwahab was inside to meet me. Getting to Paris was nerve-racking. Had to sit for five hours waiting for the plane to come in from Casablanca. Each time I asked for information, I was told that it still hadn't left the Casa airport. Never found out why. The flight was easy. It was nearly dusk when we hit Orly. Rondeau and

Claude Thomas were waiting, and had been waiting all afternoon. There were great complications in the Paris streets. Strikers had built bonfires in the middle of the avenues, there were police swarming like ants everywhere, and traffic barely moved. Finally we got out and walked, leaving my luggage in the car for the chauffeur to deposit at the hotel when he got there. John Hopkins had come over from Oxfordshire to see me, and Claude, John, and I had dinner in my room. A fine dinner—the first good steak I've had in twenty years.

Next day very busy. Bibka (Madame Merle d'Aubigné) gave a huge complex lunch for publishers and critics. I was treated like a star, and enjoyed it. The afternoon was crowded with people who came to the hotel and asked questions and took photos. Rondeau and I dined alone, and were driven to the studio. Program longer than I'd expected, but it went off easily. Pivot obviously clever; how seriously devoted to literature I don't know. He was a bit hard on Miss Siegel, although if you bill yourself as Sartre's secret mistress, you can't object to a little rough treatment.

Saturday I went shopping with Claude, hoping to find a good bathrobe. The first shop we hit had one, but it cost nearly five thousand francs, which I had no intention of spending. Finally got one of cashmere for something over three hundred dollars, which still seems rather high for a garment no one but me will ever see. But I was glad to have a trophy to bring back to Tangier. It was already night when Claude, Rondeau, and Sylvaine Pasquier said good-bye at Orly. Paris more splendid than in 1938, but I wanted to escape from it before I began to remember it.

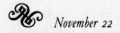 *November 22*

A telegram from Regina Weinreich says she is arriving with TV crew. I'd more or less decided that the whole idea of the film had been abandoned, since October had gone by without any word from her.

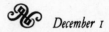 *December 1*

In an early missive Weinreich mentioned an hono-
rarium of ten thousand dollars. But on arriving she
announced that there would be no money. The
crew was here for a week. Catherine Warnow,
the *metteuse-en-scène,* kept me and the crew working
overtime. Scenes outdoors in the wind at the Café
Hafa, in the Fez Market, in my bedroom showing
me eating in bed. Hard to imagine anyone being
interested in such material.

 *December 20*

The TV crew rented Buffie's apartment downstairs
for a week, where they stored their equipment.
But according to Fatima they left it in a bad state.
(They also neglected to give her any money for
cleaning it up, so she is complaining to Steve Dia-
mond, who's now occupying the place, in the hope
that he'll tell Buffie on his return to New York.)
Steve has the flat for a month or six weeks. Each
morning he calls for me and we take the walk,

necessary for my leg, either into town or toward
the country.

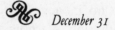 *December 31*

Last night Mrabet gave the annual birthday party
for me at his house. No orchestra, but a group of
girls who sang and beat drums. Abdelwahab hesi-
tated about going because he knows that Mrabet
dislikes him. But Mrabet dislikes everyone who
comes to see me, so I encouraged him to accept.
Great quantities of delicious food.

 *January 8, 1989*

Steve had it in his head that he wanted to give me
a parrot. When we spied an African gray opposite
the Spanish cathedral, he went into the shop to inquire the price, and was told it was two thousand
dirhams. Then yesterday we went again to look
at the bird, I all the time objecting that it was
much too expensive. The owner now wanted three

"Krazy Kat" (Kenneth Lisenbee) and Mick Jagger, Tangier, 1989 *(Credit Phillip Ramey)*

thousand dirhams, which made it possible to leave without further conversation. Steve later remarked to Abdelwahab that he thought I would rather see things go badly than well, because I seemed so relieved after we'd left the shop.

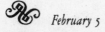 *February 5*

Three elderly people now living in Buffie's flat. Very quiet. It must be frigid down there with no heating apparatus. They brought me a copy of Buffie's book, *Lady of the Beasts.* Very impressive production.

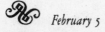 *February 25*

Books arrive practically every day from one place or another, and Abdelouahaid is indispensable in getting them through the censors and the customs, and to the postal authorities. But yesterday he came out to the car where I was waiting and in great excitement began to upbraid me. "A book that is

killing people all over the world, and you want it.
It's very bad. They're angry in the post office." I had
no idea what was the matter. "What book? Where
is it?" I got out of the car and went into the building, where I saw them all fixing me with baleful
stares. One of the employees came to me and explained. "You have a book here that's forbidden."
I asked him if I could see it, but he said it had
already been repacked, and no one could see it.
"Can't you show me the parcel so I'll know where
it came from?" He went behind a counter and held
up a package in the dark by its string, not wanting
to touch it with his hands. By this time I'd guessed
that the book was the one that was making all the
trouble, thanks to the dictator of Iran. Still I had no
idea who had sent it. Another official came up
frowning. "This is contraband goods. You cannot
have it." "I don't want it," I told him.

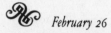 *February 26*

Today at the post office the clerks wanted to know if the police had been to see me. I said they had not. "They came here and wanted your name and address, and they took away the book to deliver it to the government in Rabat."

 March 1

The police haven't bothered me yet, so I suppose they won't. But now my incoming books are held up for an extra day while they're given a more thorough examination than they were previously.

 March 10

Another French TV show wants to send a crew next week. This one called *Ex Libris*. I've been worrying that Claude Thomas may not be translating *The Spider's House*. If she isn't by now, does that mean they'll get someone to make as bad a translation of it as they have just made of *Without Stopping*?

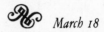 *March 18*

French TV crew here today. Interviewer, intelligent and pleasant, seemed shocked by the humble aspect of the flat, saying he'd expected me to be living on the Mountain in a big house with a fine garden. Very earnestly he said: "Do you *like* living this way?" Then he decided that the interview should be conducted in the wine bar at the Minzah, a decor more in keeping with the expectations of his public.

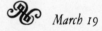 *March 19*

Interview not too lengthy, so now it's finished. To be broadcast April 5.

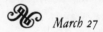 *March 27*

Young man from *Le Quotidien* came from Paris to ask questions. Couldn't give him much time.

 *March 28*

Claude Thomas arrived from Paris for a few days. She is not working on *The Spider's House.* This is the worst news yet. She's justified in insisting on having a contract with Quai Voltaire. It's something of a mystery why they keep promising the document and then fail to send it. I naturally assume that they've already given the work to someone else who will do it more quickly, since Claude is conscientious and takes her time, as indeed she should.

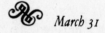 *March 31*

Went to dinner last night at Claude's, taking along Bergil Howell, who was enchanted by the beauty of the estate. The forest was dark, and the half-moon lighted the scallops of foam as the waves broke almost without sound against the rocks. The interviewer has been coming every day. It seems he's also writing a piece for the *Globe.*

"Jerez de la Frontera" (Cherie Nutting), Tangier, 1987
(Credit: Paul Bowles)

April 3

Two photographers from Rome. They had an idea
that there were picturesque cafés along the water-
front. After letting them drive all around the town
without finding anything suitable, I suggested the
Café Hafa. It delighted them. One of them had
been in Moscow last week, and he talked about
the city while the other took picture after picture,
always saying, "Look at me" between snaps.

 April 4

Suomi Lavalle came this morning and made his own
hundred or so photos, taking me up to the garden
of the Shaikha Fatima F.S. al-Sabah. (I never know
what to call her; I only know she's the daughter of
the Emir of Kuwait.) I like her very much.

April 20

Annoyed with post office. New record with eleven
of my songs, just issued, had been put on a cassette
and sent me from New York, but some employee
ripped the envelope across and removed the tape,
pasting it up again with official stickers. So no songs.
I suppose this is connected with the Salman Rushdie
book having been sent me—by Carol Ardman, who
ought to have known better, and who probably did
know better. (She wrote me last week, admitting
that it was she who was responsible.)

 April 23

Ramadan makes everything difficult. Once again
Mrabet has shown the inflaming effect it has on him.
I knew he disliked Abdelwahab from the beginning, but until Ramadan he found it possible to
behave in a normal fashion. Then, seeing that Abdelwahab was eating and smoking instead of fasting,
his dislike turned to fury. This afternoon he came
in and found A. sitting on a hassock drinking a cup

of tea. He lifted the table and flung it, tea and all, upon A., accompanying this action with a stream of invective in both Arabic and Spanish. *"Zamel! Yehudi! Maricón de mierda!"* His small daughter (to be five next week) stood by in a state of bewilderment. There will probably be more such scenes before the end of the fast, since Abdelwahab comes each noon and prepares my lunch for me, and Mrabet can't abide the thought that he should refuse to observe Ramadan and fast along with everyone else.

 *April 24*

I thought I was finished with entertaining TV crews. They've come from Milan, Amsterdam, London, Paris, and New York. But there's another crew coming, this time from Geneva. I had a German couple here yesterday and today, recording for a Berlin radio station. The woman had a penchant for beginning her questions with the word "why." I told her that questions starting with "why" couldn't

be answered intelligently or truthfully. Of course
her question then was "Why not?" I recanted and
said that my remark had to do only with me. This
didn't help, because she countered by saying: "But
we're talking only about you."

 April 25

Two disreputable-looking Moroccans rang my bell
at about two-thirty in the afternoon. They didn't
seem to know how to begin talking. Then the lift
arrived and Abdelouahaid got out of it, standing
behind them. "Your daughter wants to see you,"
said one. When I objected that I had no daughter,
he only laughed. The other said: "Yes, you have,
and she's here. The one named Catherine, from
Germany. She's never seen you, and she wants to
come and meet you." "I don't want to meet her,"
I told him. Abdelouahaid spoke up, assuring them
that there was a mistake, but they carefully paid him
no attention. "Shall we bring her here at five?"
"No, no, no! I have no daughter. Thank you, but
I don't want to see her."

They left. Abdelouahaid came in, warning me that they were criminals, and not to let them in if they came back. I felt fairly sure they wouldn't return. But after I got back from the market and post office and Abdelouahaid had gone home, the bell rang again. The two were there, looking as though they were supporting a woman between them. She wore a wide hat and kept her head down, so I could not see her face. "This is your daughter," they told me. "She comes from Essen." By this time it was all so unlikely and ridiculous that I yielded to temptation and decided to let her in, but I made the Moroccans stay outside. As if to introduce herself she pulled from her pocket a paperback copy of *So Mag Er Fallen.* Then she said in heavily accented English: "I have shame, but I go tomorrow to Germany." Apparently she could think of no way to meet me during her one afternoon in Tangier, save by going to the Zoco Chico and begging everyone who would listen to take her to her father, named Paul Bowles. The two outside the door, after arranging a price in the Zoco, had taken pity on her and claimed to know me, although surely they had never seen me.

Her conversation went in various directions and was hard to follow. I began to wonder how I could get rid of her. During the rambling she declared that she wanted to die. This made me even more eager to get her out. I gave her a cup of tea. As she drank it she explained that she had hoped to die in Merzouka on top of a big sand dune, but hadn't managed it. I said it was a pity, and she agreed. Finally she corrected herself. "I don't want to die. I want to change." Her glance was coy as she painted her lips.

I gave her a copy of *Gesang der Insekten* and signed it. She seemed disappointed to see that the locale was Latin America. It was clear that she had an obsession: she wanted to read only about North Africa. By the time I eased her out, her two friends were gone. I hope they hadn't already been paid, and were waiting downstairs, for she'd told me she had no idea of how to get back to her *pension*.

April 26

A house down the street has a large stork's nest on top of its chimney. Each spring a family of birds arrives, stays two months or so, and then goes on. Last year there were two young ones. They moved around the nest constantly and practiced flying by jumping up and down and flapping their wings. The male, apparently annoyed by all the hubbub, built himself another nest on the top of an electrical pylon about a hundred feet from the first. During the winter workmen came and pulled down the nest. Yesterday I noticed that the storks were in residence. Once again a big nest has been built atop the tower. Is it the same couple each year, and do the young birds return with their parents?

I haven't seen storks migrating for many years—thirty at least. I used to go down to Merkala and see hundreds of them moving past in perfect V-formations, so low that I could hear the regular beat of their wings. In the spring they flew out across the strait toward Spain, and in the autumn they came back. Storks strike me as particularly

Bachir el-Attar, Jajouka, 1990
(Credit: Cherie Nutting)

beautiful in flight, in spite of the two sticklike legs
that dangle beneath. The long neck and the great
wings slowly beating are what one notices.

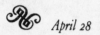 *April 28*

Scarcely an afternoon passes without a visit from
someone I never saw before and probably shall
never see again. Giving all this time makes life seem
a static thing, as though an infinite number of years
lay ahead. It's become a serious problem only during the past year or two. Not having a telephone
makes it worse: people come all the way out here
and knock on the door. This makes it difficult to
refuse to see them. Someone has come every afternoon this week.

 May 10

Sometimes I wonder how long this routine is going
to continue, the eating of all meals in bed. After the
sympathectomy the doctor thought it a good idea.

Now, after nearly three years, I go right on. At noon Abdelwahab comes and makes my lunch. In the afternoon Mrabet brings food and prepares dinner. Breakfast is the most important of the three meals, but Mrabet doesn't seem to understand this, since recently he fails to appear in the morning, so that I have to invent my own breakfast and eat it at the kitchen table. I consider myself lucky to have friends willing to serve me in bed. But Abdelouahaid says: "You're not sick. Why do you have to eat in bed?" Of course, Moroccans have a horror of staying in bed, even if they're running a fever. Anyone who takes to his bed with an illness is presumed ready to die.

Someone came from New York and presented me with a copy of that biography Sawyer-Lauçanno insisted on writing, even though I begged him not to. His annoyance with me for refusing to cooperate shows clearly, so that it becomes, whether or not he wishes it to, a defamatory work. In order to produce it he was obliged to use my autobiography. The difference between mine and his lies in his decision to doubt the veracity of my account and substitute his own version. Doubtless he thought his own more piquant since it makes me out to be a liar. There's an endless list of false statements based purely on assumptions and hearsay, but with obvious malevolence.

For one thing, Sawyer-Lauçanno was much too ready to listen to certain mischievous gossips, and to include their scurrilous nonsense in his text, without bothering to verify. That's annoying, but at least the so-called information is credited to those who proffered it. What is infuriating is his tendency to imagine that I gave what I knew were false versions of events. The truth is too tame for his kind of journalism. I'm angered by his portrait of me as an accomplished moocher in the thirties. It's enraging to read his personal story about why Jane left Portugal in 1958. She and I went every day for two weeks to the American Embassy in Lisbon to procure a new passport, hers having expired while we were in Madeira. "No," they said. "We must get in touch with the FBI first." We continued to go to the embassy, in the hope of a quick reply from Washington. It was not quick, and when it came, it was negative. "Mrs. Bowles will have to return to the United States immediately. We'll give her a paper in place of her passport." Jane left a day or two later. Mr. S.-L. says that the State Department assured him that such proceedings were quite impossible, and so he assumes that my story is fictitious and that I used it as a "ruse" in order to get Jane off my hands. The book is defamatory, but unfortunately not actionable.

 May 12

Many letters from Jerez in New York. Bachir is with her. She speaks of arranging jobs for him. One of these was the La Mama production of *The Night Before Thinking,* the play that Ahmed Yacoubi had hoped to see produced several years ago, and which I'd vetoed because his collaborator had excised the baby with eyes everywhere on its body. The baby seemed to me the most important character in the story. Now the piece was presented (without baby, of course) and Bachir played *rhaïta* and *liara* and even *guinbri,* but he was annoyed by La Mama; finally Jerez decided that La Mama was crazy.

In one letter Jerez said that she and Bachir had been married "at the mosque around the corner." This will make it easier for him to extend his stay in the U.S., important because he hopes to go on tour with various groups, including The Rolling Stones.

 May 25

Instead of an American tour, Jerez managed to arrange a Moroccan trip during which Mick Jagger would visit Jajouka with a BBC crew, to perform with Bachir in his native village. Jerez and Bachir will arrive here June 4, and the Stones a few days later.

June 2

Phillip and Krazy Kat just in from New York,
Phillip laden with gifts. All that was lacking here
in the flat was a Christmas tree. Krazy Kat is
pleased that he'll be meeting Jagger. I of course
am pessimistic about the whole project, suspecting
that the BBC won't have had time to prepare the
necessary formalities for getting equipment into
the country. I remember how the Dutch had to
make three trips here from Holland before they
were able to persuade the officials to allow their
recording equipment through customs.

Some of Phillip's most important gifts were the
new records, one of *Music for a Farce* (Reference
Recordings, Chicago Pro Musica) and the other of
songs sung by William Sharp, who gives what I'd
call definitive performances of eleven of my old
songs. The *Farce* pieces sound very fresh and imme-
diate, after years of hearing the old Columbia and
MGM recordings.

June 3

I wrote Sawyer-Lauçanno, giving my outraged reaction to the voicing of all his suspicions. Then he replies: "At the urging of my editor, I made an inquiry to the State Department. They categorically, even emphatically, denied that they would have done such a thing. . . . In retrospect I can see how I created the impression that I did not entirely believe your account. All I was trying to do was give another version of the story." He should have displayed his inventive virtuosity by giving three or four more versions.

 June 5

Jerez and Bachir are here in Tangier. Both are excited about the Stones' arrival, and I don't dare voice my pessimism. They have to be careful to come to see me only when they're reasonably sure that Mrabet won't be here.

Now that Bachir is Jerez's husband, he naturally feels that the house built with her money

Paul Bowles, Tangier medina, 1987
(Credit: Phillip Ramey)

belongs partly to him. I think he knows that there's virtually no chance of his ever getting it, since it's registered in the name of Zohra and not of Mrabet.

 June 7

Some advance crewmen from the BBC arrived with a view to shooting Jagger when he comes. *If* he comes, I thought. Their chief seemed pretty sure he would.

 June 8

Today everyone is much less certain. Customs refuses to let any of the BBC equipment into the country. Great *Geschrei,* and much telephoning to London. Ultimatum from Stones' manager: If equipment is not through customs by nine o'clock tomorrow morning he will cancel the engagement. Jerez arrived this evening, panting with anxiety. "You've got to call the king," she told me, although she knows I've never met him and wouldn't call him even if I had. Then she mentioned that she'd been to see Lalla Fatima Zohra during the day and had been told to call her by telephone later. Her eyes lighted on me. "*You* know Lalla Fatima Zohra. You call her. Tell her how important it is for Morocco to have BBC make this film." I was in my bathrobe, downstairs with the others, in Buffie's bedroom. Everyone seemed to be of the opinion that the least I could do was to call Lalla Fatima Zohra and explain the situation, which I think Jerez had found it difficult to do because of language obstacles. (I had forgotten that Lalla Fatima Zohra speaks perfect English.) "But I can't speak Arabic well enough to talk to her," I objected. Abdelwahab made a suggestion that I call her and give my name, then pass the telephone to him. This we did, and she asked Abdelwahab to call back in a half hour. When he made the connection later, she seemed to be speaking into two telephones at once. There were very long waits, while everyone looked at Abdelwahab to see by his facial reactions what words were coming through from the palace of Moulay

Abdelaziz. He merely said, *"Naam, lalla,"* from time to time. The conversation, if that's what it was, went on for ten minutes. At the end Abdelwahab hung up, saying that she had promised to call customs and ask that the television equipment be let in. By that time it looked as if The Rolling Stones would not be coming to Tangier. I said that I doubted Lalla Fatima Zohra had the power to force the airport officials to do anything, one way or the other. Both Bachir and Abdelwahab hotly disputed this. "Her word is law in Tangier," Abdelwahab cried. I said I certainly hoped so, and came upstairs to sleep.

 June 8

Abdelwahab, who is almost as interested as Bachir in seeing that the film gets shot, came by at noon to tell me that Jagger and Keith Richards had gone to the Intercontinental and refused the suites reserved for them by Jerez. (Abdelwahab is getting married next month and is trying to have as good a time as possible before the wedding.)

So permission was given and the Stones came from London. All day yesterday and today Jerez, never once doubting that all would be well, has been rushing from one place to another, trying to find a likely place where the film could be shot. She has no time to eat, can't sleep at all, and will be ill in another day or two. She got a go-ahead from Malcolm Forbes for equipment to be set up in his garden, but the TV people turned it down in favor of the courtyard of the Akaboun house, which I suggested because it provides a more authentic background. About five o'clock Jagger arrived, accompanied by so many others that the room was *archicomplete.* Keith Richards was among them. He paid his respects to me and left, saying he was going to bed. Jagger sat down beside me, and we started to talk. It was a while before I noticed that our conversation was being filmed. After a quarter of an hour the filming stopped. "I'm tired," he said. "My kids woke me up at daybreak today. You see, Sunday's Father's Day, and they had to give me their presents today before I left. See you tomorrow at the show."

June 10

Sat for two hours this afternoon in the Akaboun
courtyard. There were sixteen men from Jajouka
with their *rhaïtas* and *benadir,* all wearing heavy
brown djellabas. Much too hot for all that wool.
They were impressive and sounded magnificent.
How they sounded on film God knows, but it probably doesn't matter. Bachir did a lot of solo work
with and without drums. The only sounds made by
Mick Jagger were made on drums. He may have
sung after I left, but I doubt it. In the adjacent room,
whenever there was no music, I could hear a kind
of pedal point that went on without respite, a constant tonic, very low in pitch, and not quite a
quarter-tone higher than the tonic of the key in
which the *rhaïtas* and the *liara* played.

 June 12

Had lunch on the Mountain with Gloria Kirby. Her
guests from Madrid knew all about Pedro Almodóvar, who seems to specialize in comic films. It's hard

to understand, if that's the case, why he chose to take an option on *The Time of Friendship,* a story in which there's not a suggestion of a humorous situation. Unless he does the whole thing tongue-in-cheek. It wouldn't be difficult to satirize the crèche scene. Or making Slimane three or four years older could provide a different sort of liaison between him and Fraülein Windling. But humor? The guests at lunch suggested that Almodóvar felt he had exhausted his comic vein, and intended to add a serious dimension to his work.

 June 16

Yesterday the American ambassador to Rabat set up an appointment with me at the Minzah bar. Seemed pleasant enough, but I had no idea why he wanted to see me, and still have none.

June 20

After lunch Rodrigo, as he often does on fine days, went to Merkala in order to swim out at the Sindouq. Before my artery got blocked I used to go almost every day, following the coast, climbing among the boulders, jumping from one to the other. It was my favorite activity, imitating a chamois, sure of my feet. If someone mentioned that I was like a goat, I joked about my zodiacal sign. "Of course, Capricorn." I can hardly get along the shore now even if I have someone pulling and pushing. Pointless to pretend that time makes no difference.

 June 24

Last night Bertolucci sent a car for me, to take me to the Minzah for dinner. At the beginning of the meal he said: "At last, it's happening." "Yes. For two years I've been wondering whether it would," I told him. Everyone connected with the making of the film was there, including the producer, whom

Paul Bowles, Café Hafa, Tangier, 1987
(Credit: Phillip Ramey)

I'd met a few years ago when Bill Burroughs was
here with him from London. Conversation was difficult. A very noisy floor show was going on for the
benefit of a huge group of shrieking tourists. Bertolucci brought up the subject of music. He was still
thinking of using David Byrne, although he mentioned Richard Horowitz as well and at one point
said he'd like me to provide some of it. We didn't
discuss it. I suspect he'd like electronic material
rather than symphonic. Much easier, much cheaper.
No parts or rehearsals needed. Scarfiotti had mentioned that he'd like to use Agadèz as the setting for
the final city in the south. I hope this can be
managed, and that they don't try to shoot everything in Morocco. I can appreciate their not wanting
to get involved with the Algerians, but Morocco is
no substitute for Algeria or Niger.

 June 30

Yesterday a French TV crew came to interview. Nothing memorable. We repaired to the

belvedere at Sidi Amar, and they shot a long scene there against the sea. Occasionally I can view these things on video afterward, but not usually. They're disappointing for always coming out in black-and-white.

 July 5

Yesterday lunch at Gavin Young's, taking along Abdelwahab. We had mint tea in the garden afterward. When I picked up one of the pillows off the lawn, I found a very excited centipede on it. Abdelwahab: "Don't kill it! Please!" Gavin, later: "Have you become a Buddhist?"

At noon today I had a visit from Riki Suzuki, who presented me with a copy of my new story collection from Shinchosha in Tokyo. As usual, I was a little ill-at-ease, as I always am with the Japanese. It's so hard to know what they're thinking, and to know how awkward I'm being, according to their standards. Are they in the act of approving, or disapproving?

July 17

A Moroccan sent from Holland by Robert Briatte
arrived this morning. I can't quite make out what
he wants from me. I know he hopes to make a
film, but since he's leaving for the U.S. shortly, it
doesn't seem likely that he'll do it very soon. We
went, at his insistence, to Café Hafa. On the way
Abdelwahab, with his brother-in-law, overtook us
in a car, and drove us the rest of the way to the
café.

Another Moroccan came in the afternoon—a
professor at the University of Limoges. He was
quite frank about his intentions, and pulled out a
tape recorder immediately.

 July 19

Today Gavin asked Abdelwahab and me to lunch.
He's about to leave for England, but will be back
here in October, when he'll tell me more about
being designated as a chief in British Samoa, a func-
tion he expects to assume this coming winter. He

suggested I go out there and see him in regalia, knowing I shan't.

 July 21

There are three Moroccans coming regularly to ask me questions: Mohand, Rais, and Fqih Aouami. When Rais arrived the first time, he brought roses and a sugar loaf. As he handed the loaf to me he said: "You know the tradition." I agreed that I did, which was untrue. I still don't know, because no one seems to be able to explain it. Fqih Aouami is a professor at the University of Limoges, and has many questions prepared.

 July 25

Last night Philip, Krazy Kat, Rodrigo, and Lidia Breda were with me at Abdelwahab's wedding. Chaotic and noisy. More than a hundred guests. The music was *raï,* deafening. It made conversation almost impossible, and dancing almost obligatory.

Only the men and boys danced, moving in violent gyrations hour after hour without appearing to tire. The girls sat in long rows, watching.

 July 27

Abdelwahab tells me that the family of his bride came and demanded of his parents that they give them the nuptial sheet stained with blood. The bridal couple, after forty-eight hours of celebration with no sleep, had merely collapsed on the bed and lain inert for a two-hour respite before being called to continue the festivities. Thus there was no blood. Abdelwahab's parents were outraged. "Such backward people!" they commented, but they did agree to let the bride's family have the sheet once it was marked with blood, which presumably will be tomorrow.

July 28

Last night Fatima es Sabah sent a car for me, and I
was able to attend her garden party, where Bachir
and a group of Jajouka musicians performed sitting
in a line atop a big boulder above the pool. As the
guests went down the path to the garden, they
passed a row of ten servants in livery standing side
by side, to greet them. The princess, however, was
not in evidence, and David Herbert took great exception to her absence, exclaiming: "This . . . is not
my way." Previously, up at the house, she had
brought out a handsome camel's-hair garment, a
Kuwaiti variation on a djellaba, for me to wear,
because she knew I was *frileux,* and the garden,
being at the edge of a high cliff over the ocean, is
subject to sudden bouts of chilly wind. The cape was
welcome. I spent a certain amount of time being
photographed by Suomi Lavalle lying on the rock at
the feet of the musicians, and without my covering
of camel's hair I'd have been cold. When I left,
Fatima insisted that I keep it because, according to
her, it suited me perfectly.

July 30

The Yarmolinskys gave a Jilala party last night at Villa Julie. There was dancing, especially by women, some of whom seemed to be in a state of trance, although I don't believe they were. (Mejdoubi's widow was particularly pleasing.) The couscous was excellent. I was fortunate to be eating inside, for there were mosquitoes in the garden. Abdelouahaid amused us by calling for us in a huge bus that held seventy passengers. He played the clown, driving us up the mountain and then down again, ostensibly without brakes. Phillip was worried that he'd go over the edge of the road, but of course he knew what he was doing and got us all home without a mishap.

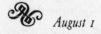 *August 1*

Dinner last night at Abdelwahab's, with his bride. Phillip, Krazy Kat, Rodrigo, Lidia, and I watched two video clips of life with P.B. One was Gary Conklin's old film which was recently shown on

French TV, and the other was a part of *Ex Libris*—
the least interesting part. We were also entertained
by endless sequences of last week's wedding. Dinner was good, and we left after Abdelouahaid had
agreed to come into the house and sit, despite his
feud with Abdelwahab. Farewells were lengthy and
emotional. The bridal couple leave this morning for
the Netherlands. I'll miss Abdelwahab.

 August 2

Went tonight to the Marquis for dinner with Gold-
stone. Food not as good as usual. He wants Mrabet
to consult his cardiologist for a checkup and insists
on paying for the consultation.

 *August 7*

Went last night to the annual Callaway dinner at the
Marquis. Roast beef a little tough, but better than
one would expect to get in Tangier, where beef is
invariably awful.

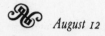 *August 12*

Now I must take back what I said on Monday about beef in this city. Last night Phillip gave a dinner at a restaurant where I'd never eaten, called Osso Buco. I had a filet that was perfect, which I'd have considered perfect in Paris or New York. Why that restaurant should be able to get good beef while all the rest is nearly inedible, I don't understand. Jerez was there. Bachir came in unexpectedly while we were eating and sat down with us to drink beer. He and Jerez were not on speaking terms.

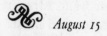 *August 15*

The Moroccans keep coming every day. Sometimes it's hard to keep track of what each one wants. Fqih Aouami is the most difficult one to satisfy; he always has a new set of questions. Last night I heard drums—not the *darboukas* of Aachor, but in a variety of timbres. I had the impression that an *ahouache* was in progress. Never having witnessed an *ahouache* save in the High or

Anti-Atlas, I decided that my hearing was defi-
cient. When I opened my bedroom window and
heard the chanting above the drums, I was no longer in any doubt. An *ahouache* was going on, up
the street in the vicinity of the school. I ran out
and found thirty men in traditional white robes,
each with his dagger, dancing in a long line. Eight
drummers crouched in front of them. I might have
been in Tafraout. I stood motionless for about an
hour, mystified and delighted, until they filed out
of the courtyard. Then I asked a policeman sitting
at the gate how it happened that such a group
found itself here in Tangier. "They were brought
by the American chief," he told me. "You mean
the government?" "You know, the chief with the
palace on the Marshan." That could mean only
Malcolm Forbes. I have an invitation to a dinner
he is giving on the nineteenth. A great idea, I
thought, to bring performers from the deep south
all the way here. There were nine very large buses
parked outside the entrance to the school, which
means that several hundred dancers and singers
have come.

August 16

I went again last night to watch the performers, but
arrived too late to see anything but a few youths
dancing. They were good, but I don't think profes-
sional.

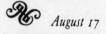 *August 17*

Jerez and Bachir came last night with the idea of
going to the school with me. This time there was a
crowd of two or three hundred men dancing, and
perhaps fifty women. A great performance, including an astounding group of Haouara and even
Bechara with her girls, dancing to the *guedra*. Bachir, who used to work in Goulimine and knew
Bechara from that time, spoke with her after the
rehearsal. She asked us to have tea with her upstairs
in a room devoted to the *guedra* dancers. When she
saw me she claimed to remember me, although I
don't quite see how she could. The last time I saw
her was in 1962 backstage in New York. (Kather-
ine Dunham had imported her and three dozen

other Moroccans to appear in her ill-fated musical;
it lasted two nights.) I'd recorded her in Goulimine
in 1959, but she wouldn't have been likely to have
recalled that one evening. In any case, we were
received with the traditional Moroccan hospitality.
Some of the dancers reached into their bosoms and
pulled out bundles of silver trinkets, which Jerez
and Bachir bought. When we left, Bachir told me
she had said to him: "If that American lives in Tangier, he must have a lot of money. Is he married?"
We agreed to go back tonight and see her.

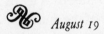 *August 19*

Last night we did go back, but everyone was out at
the airport to welcome the planes arriving with
Forbes's guests from New York. It must have been
an impressive scene, with all those members of the
Royal Guard on their black horses, and the long
lines of dancers in white robes and turbans. On the
airstrip there would have been ample space for everything, whereas later in the street it would

necessarily be crowded. I'd like to have seen it, but security was so tight that no one could have gotten into the airport to see anything.

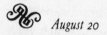 *August 20*

The party must be over by now. I'm told that it went on all night. By midnight I'd had enough. There remained only the birthday cake and the fireworks to look forward to, and I saw the pyrotechnical display from my bedroom window after I got home. A tiring evening. First, the police refused to let us anywhere within a quarter of a mile of the entrance, so that we had to park in the *plazuela* behind the Café Hafa. The crowd in the street was so dense that Abdelouahaid had to push people aside roughly to make it possible for me to make any headway. When we got within sight of the Palais Mendoub, he wished me good luck and went back. There was a long queue of guests waiting outside in the street. As we moved slowly forward we were pelted with rose petals by girls standing on each side of the queue. Beyond were the ranks of the dancers and drummers, and in the background, along the wall of the garden of the Palais du Marshan, stood the horses with their uniformed mounts, still as statues. I could think only of how fortunate it was that the weather was fine. Even a few drops of rain would have ruined a hundred evening dresses. It didn't seem an ideal manner of welcoming guests, to force them to stand in the street for a half-hour waiting to get to the bottleneck just outside the gate. There we exhibited our invitations, had our names checked on lists, and were admitted one at a time. The line continued through the courtyard, until we were given maps of the terrain and assigned to our tents. I counted nine of these objects, once I had passed through the receiving area, where our host stood grinning, flanked by his sons, and with Elizabeth Taylor seated by his side. "Wait till you see how fat she's grown!" people had warned me. To me she didn't look fat; she looked solid and luscious. She must have been tired; it's not easy to be introduced to nine hundred people one after the other. I refused the champagne and set out in search of the tent I'd been assigned to, pushing my way through the crowd until I'd found it. There were no place-cards. I sat down at an empty table until a waiter asked me to choose another, also empty. No one seemed to be in a hurry to eat. My table did eventually fill up—with, among others, the governor, the chief of police, and a military man decked with medals. At the next table sat Malcolm Forbes and his family. Miss Taylor was on his left, her back to me. The crown prince sat on his right. For three hours as I ate I watched their table. The French woman next to me made repeated comments in a whisper about Elizabeth Taylor's shoulders and the crown prince's face, which she characterized as "frightening" and "almost Japanese." All I could reply was that he never altered his poker-face expression and spoke very little. I myself thought he was unutterably bored; if that is so, it was understandable.

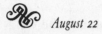 *August 22*

People who did not attend the Forbes dinner ask: "What was it like, really? Was it a ghastly farce?" Nothing so frankly commercial can properly be called a farce. It was what sixty years ago would have been called "spectacular and colossal," the difference being that in this case the adjectives are apposite.

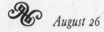 *August 26*

A man named Jancovici who runs Les Éditions de la Différence in Paris has been bringing me art books (very well produced) in the hope that I'll provide a text for a book of reproductions of paintings by a *Mallorquín* named Barceló. I'm still busy on the piece for the Munich magazine, and trying to prepare the third part of this *Journal Tangérois* for Briatte and Librairie Plon. It occurred to me that it might be possible to do the Barceló book as fiction rather than as exposition. The paintings are pale gouaches, mainly monochromatic. If I were to

illustrate them with a story of some sort, the weld-
ing of the two elements might stick.

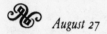 *August 27*

Bertolucci now thinks I should appear in certain
scenes of the film. I don't understand exactly why,
and therefore suspect this to be a whim which he'll
possibly think better of sooner or later. Richard
Horowitz is busy gathering material to use in his
soundtrack; I hope he doesn't decide that Moroccan
music will make a satisfactory auditory backdrop for
the Algerian Sahara.

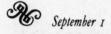 *September 1*

Buffie left today for New York. We had lunch yesterday in the garden at Güitta's, after which she
said: "She's lost a customer," referring to Mer-
cedes, who runs the place single-handed. The sole
was far from fresh. There was too much insect life
around: flies by the dozen, ants hurrying along the

tablecloth, and an insistent hornet which crawled on my fish and nibbled on it voraciously throughout that entire course. In the afternoon went with Umberto Pasti (of the Italian *Vogue* piece) and his associates to the Avenida de España for a lot of photographs.

 *September 4*

The king's ship was due to arrive back from Tripoli today, so everything was shut, the market padlocked, the streets barricaded, so that in order to get to the Almohades, Abdelouahaid and I had to walk in various directions through mud and piles of gravel. (The city is a mess, with big apartment houses going up everywhere.) At the Almohades, where the Sahara Company's office is located, I was measured for the film's costume designer, who is in London at the moment. What sort of clothes is he going to make for me, and in which scenes will I wear them?

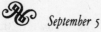 *September 5*

The thousands of people waiting along the water-front to see the king's ship come in waited in vain throughout the day. Last night, during a very loud and protracted thunderstorm, the vessel arrived, and the king was spirited aboard his train for Rabat. Neither he nor anyone in the government trusts the people of Tangier, and so he makes a point of not ever coming here if he can help it. I've never understood the official antipathy for Tangier and Tetuan, but doubtless those who feel it have their reasons. I returned to the Almohades this afternoon with a message for Bertolucci; while I was there he called from the Tafilelt, where he said the heat was intense. He told me he was en route to Algeria tomorrow: Béni-Abbès. It will be hotter there.

PEOPLE NAMED IN
THE TEXT

BACHIR EL-ATTAR: Moroccan musician, leader of the Master Musicians of Jajouka, performer on the *liara, rhaïta,* and *guinbri* who has recorded with Mick Jagger and The Rolling Stones.

"JEREZ DE LA FRONTERA" (CHERIE NUTTING): American photographer married to Bachir el-Attar, whose work has appeared in the *Boston Sunday Globe Magazine* and the Swiss magazine *Du,* among other publications.

BUFFIE JOHNSON: American painter and author of *Lady of the Beasts,* a study of the female element in the history of religions.

GAVIN LAMBERT: British-born author of novels, stories, and screenplays, whose most recent book is a biography of actress Norma Shearer.

MOHAMMED MRABET: Moroccan storyteller, author of numerous books translated into English by Paul Bowles, including the novels *The Lemon* and *Love with a Few Hairs* and an autobiography, *Look and Move On.*

PHILLIP RAMEY: American composer and writer, longtime annotator and program editor of the New York Philharmonic.

RODRIGO REY ROSA: Guatemalan writer whose short-story collections *The Beggar's Knife* and *Dust on Her Tongue* and novel *The Pelcari Project* have been translated into English by Paul Bowles.

GAVIN YOUNG: former foreign correspondent for *The Observer,* among whose books are the celebrated *Slow Boats to China* and *Slow Boats Home.*

PRINCESS LALLA FATIMA ZOHRA: daughter of the Sultan Moulay Abdel Aziz and cousin of Hassan II, the present King of Morocco.

BOOKS BY
PAUL BOWLES

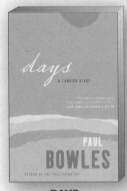

DAYS
A Tangier Diary
ISBN 0-06-113736-7
(paperback)

**THE DELICATE
PREY**
And Other Stories
ISBN 0-06-113734-0
(paperback)

**A DISTANT
EPISODE**
The Selected Stories
ISBN 0-06-113738-3
(paperback)

**LET IT COME
DOWN**
A Novel
ISBN 0-06-113739-1
(paperback)

POINTS IN TIME
Tales From Morocco
ISBN 0-06-113963-7
(paperback)

**THE SHELTERING
SKY**
A Novel
ISBN 0-06-083482-X
(paperback)

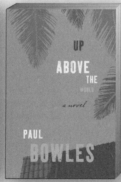